Caroline Hwang

photography by Julia Stotz

PROBIOTICS

hardie grant books

CONTENTS

Introduction   6
Ingredients & Equipment   8
Rules of Fermentation   10
Resting Solutions   12
Troubleshooting   13

Kefirs   14
Basic Water Kefir   16
Basic Coconut Water Kefir   18
Basic Milk (or Coconut
   Milk) Kefir   20
Orange-sicle Water Kefir   22
Strawberry Ginger Water Kefir   24
Pineapple Cucumber Water Kefir   26
Watermelon Mint Water Kefir   28
Pink Grapefruit Water Kefir   30
Lime Lemongrass
   Coconut Water Kefir   32
Passion Fruit
   Coconut Water Kefir   34
Cucumber Mint
   Coconut Water Kefir   36
Ginger Turmeric
   Coconut Water Kefir   38
Breakfast Milk Kefir Shake   40

Mint Cacao Milk Kefir Shake   42
Café au Lait Milk Kefir Smoothie   44
Peanut Butter & Cacao Milk Kefir   46
Piña Colada Coconut Milk
   Kefir Shake   48
Mango Lassi Coconut Milk
   Kefir Shake   50
Blueberry Coconut Milk
   Kefir Shake   52
Cashew Cinnamon
   Coconut Milk Kefir   54

Kombucha   56
Basic Kombucha   58
Hibiscus & Lime Kombucha   60
Lemon Verbena Oolong Kombucha 62
Apple Cider Kombucha   64
Winter Citrus Kombucha   66
Earl Grey & Lavender Kombucha   68
Watermelon Jalapeño Kombucha   70
Carrot Orange Kombucha   72
Pear & Spice Kombucha   74
Herbal Kombucha   76
Plum Darjeeling Kombucha   78

Jun   80

Basic Jun   82
Jun with Jasmine Tea & Peach   84
Jun with Mixed Berries   86
Jun with Pink Lady Apples   88
Jun with Strawberries & Thyme   90
Jun with Lavender & Blueberries   92
White Jun with Elderflowers   94
Jun with Blackberry & Lemon Balm 96
Jun with Grapefruit & Cardamom   98
Jun with Blueberry & Vanilla   100

Nature's Fermentation   102

Basic Beetroot Kvass   104
Basic Pineapple Tepache   106
Basic Ginger Bug   108
Spiced Beetroot Kvass   110
Orange Ginger Beetroot Kvass   112
Pomegranate Beetroot Kvass   114
Blackberry Vanilla Tepache   116
Pineapple Black Pepper Tepache   118
Lemon & Lavender
   Pineapple Tepache   120
Ginger Beer   122
Ginger Lime Mint Soda   124

Pineapple Ginger Soda   126
Persimmon & Vanilla Ginger Soda   128
Strawberry Shrub   130
Concord Grape Shrub   132
Pear Cinnamon Shrub   134
Peach Champagne Shrub   136
Tangerine Rice Vinegar Shrub   138

Whey-fermented Soda   140

Basic Whey   142
Whey Lacto-fermented Root Beer   144
Whey Lacto-fermented Lemon Soda 146
Lacto-fermented Orange
   Vanilla Soda   148
Lacto-fermented Rosewater   150
Lacto-fermented Summer
   Melon Soda   152
Lacto-fermented Grape
   & Mint Soda   154
Lacto-fermented Plum
   & Thyme Soda   156

Index   158
Acknowledgements   160

INTRODUCTION

People say happiness and health start in the gut, and studies have shown that this is true. The health of the gut is linked to alleviating stress and anxiety, reducing skin problems, boosting the immune system and more. In addition to keeping a healthy diet and being active, you can aid your gut health by consuming probiotic foods and drinks that have been fermented. What are probiotics, you may ask? Probiotics are live cultures and bacteria that help keep the gut healthy.

Probiotic drinks such as kombucha and kefirs are becoming ever more popular in the world of health drinks, coming in very close to cold-pressed juices and smoothies. Lesser known probiotic drinks like beetroot kvass, jun and tepache all have the same fermented goodness, not to mention bringing delight to the taste buds. All of these fermented drinks require patience but can be made simply. Some fermentations require the aid of living cultures like grains and scobys (symbolic cuture of bacteria and yeast) while others ferment all on their own – these are called wild fermentations. Each fermentation requires a little tending and care, but nothing more than 5 minutes of your day.

This book is divided into five chapters exploring both fermented drinks made with cultures and wild fermented drinks. Each chapter starts with a basic recipe and instructions for the foundation drink followed by different variations based on that drink. The fizzy delights will soon be tickling your taste buds, not to mention bringing joy and health to your gut.

INGREDIENTS & EQUIPMENT

Ingredients

 Water kefir grains – small, translucent, gelatinous grains used to make water kefir. They contain beneficial bacteria and yeasts that break down sugars in the fermentation process to create probiotics. They can be used activated (soaked) or unactivated.

 Milk kefir grains – similar to water kefir grains with the exception of their appearance. They are both gelatinous but milk kefir is white. Like water kefir grains, they contain beneficial bacteria and yeasts that feed off milk.

 Scoby – this stands for Symbiotic Culture of Bacteria and Yeast. It looks like a gelatinous, flat, translucent pancake and is used when making kombucha. The scoby metabolises the sugar and tea to ferment and transform into kombucha. The original (mother) scoby will eventually start producing new (baby) scobys with each batch of kombucha. You can begin a scoby hotel where you collect baby scobys and store them in an airtight container with enough kombucha to cover. Otherwise you can compost, give away or throw away your baby scobys.

 Jun scoby – this is similar to a regular scoby, except it cultures in green tea and raw honey to ferment rather than black tea and sugar and is smaller in size. Jun scoby creates a less acidic probiotic drink compared to kombucha. Similar to a regular scoby, it will create baby scobys as well. Follow the basic instructions for what to do with your baby jun scoby as you would a kombucha scoby.

 Sugar (preferably organic and unrefined like sucanat/raw sugar) – yeast and bacteria feed on sugars to ferment and create probiotic drinks. It creates lactic acid and carbon dioxide, which ferments and creates carbonation in the drinks. Honey is used in jun rather than sugar.

 Fruit, herbs, spices – different fruits, herbs and spices are used to flavour all the drinks featured in this book in the second fermentation process. Don't feel limited to what is suggested, feel free to experiment with your own version of these drinks.

Milk – to make milk kefir, milk supplies the natural sugars that the kefir grains feed off.

Coconut milk – a dairy-free version of milk kefir can be made with coconut milk.

Water – when making any probiotic drinks, use spring and filtered water only. Any chlorinated or unfiltered water can stunt the growth of good bacteria in your ferments.

Tea – this is required to make kombucha and jun, and is mainly what the scoby feeds on. It is needed for its minerals and nitrogen. While you can use herbal teas or decaffeinated teas, eventually your scoby will die due to the lack of nutrients it needs. Try doing a couple batches of decaf or herbal teas and then going back to a caffeinated tea to feed your scoby.

Equipment

Bottles and jars – in order to begin fermentation you'll need large 1-litre jars to start the fermentation. Bottles or jars with tight-fitting lids are used in the second fermentation where additional flavours are added and the carbonation builds.

Muslin – certain fermentations like kombucha, jun and kefirs require oxygen for the yeast and other micro-organisms to reproduce and ferment. These are called aerobic fermentations. You'll use muslin to cover the jars but let oxygen in and keep any fruit flies out. Muslin is also needed to strain out the whey from yoghurt to make whey lacto-fermented sodas.

Non-reactive (nylon or stainless steel) sieve – a sieve is required to strain out kefirs, scobys, tea leaves and other fruit, herbs or spices used to flavour your probiotic drinks. Scobys and kefirs are reactive to metal so a non-reactive mesh is required.

Funnel – a funnel is needed when transferring your first ferment to a bottle for a second ferment. It'll help with the spills!

Blender or food processor – to make smoothies out of milk kefirs and to purée fruit to flavour your water kefirs.

Marker and labelling tape – to label and date your ferments. Most ferments take anywhere from 2–5 days, sometimes longer if the temperature in your house is cold.

RULES OF FERMENTATION

Use the tips and advice in this rundown of basic rules to guide you through fermenting your probiotic drinks.

1. Cleanliness

Always use clean hands and clean jars/bottles. You don't need to sterilise your jars and bottles but make sure that they're cleaned with hot water and soap.

2. There are two main kinds of fermentation for probiotic drinks

Cultured fermentation uses a starter culture like kefir grains or a scoby to start a fermentation process. You can find these items online or from a friend who has been fermenting their own drinks.

Wild fermentation is a type of fermentation that uses natural yeasts and bacteria in the air combined with the natural sugars in fruit, fruit and vegetable skins or roots to start the fermentation process. This includes beetroot kvass, pineapple tepache and ginger bug, which are all covered in this book.

3. Starting from scratch

You will need a scoby to be able to start kombucha and jun drinks for the first time, which you would normally get from a previous ferment. However, if you are starting from scratch, you can purchase a kombucha scoby or jun scoby online or you could try obtaining one from a fellow fermenting friend. When you purchase a scoby it comes with starter tea. If you're obtaining from a friend, you should ask for 120 ml starter tea, which the baby scoby will rest in.

4. To cover or not to cover?

There are several different kinds of fermentation but the main two are covered in this book. Aerobic fermentation requires oxygen during the first stage of fermentation, these include jun, kefir and kombucha. Cover these ferments with a muslin to allow oxygen to circulate.

The other fermentation is anaerobic, which doesn't require any oxygen and therefore the vessels can be covered with a tight-fitting lid. Beetroot kvass ferments via an anaerobic fermentation.

5. Different stages of fermentation

When fermenting drinks, there are usually two stages of fermentation. The first stage uses cultures and basic ingredients. The second stage is for adding additional flavours and variations. The second stage is also when carbonation begins to build in a bottle or jar with a tight-fitting lid.

6. Temperature and sunlight

The temperature in your house has a direct effect on your fermentation. If it's too cold, it may take longer to ferment. If it's too hot, it may only take 1–2 days to ferment. Check on your batch accordingly. Don't let the fermenting batches sit in the sun; always put them in a dark place.

7. Carbonation and 'burping'

Carbonation naturally occurs in most probiotic drinks, with the exception of milk kefir and beetroot kvass. Carbonation means that the sugars have converted to carbon dioxide and is a sign that fermentation is occuring. The fizziness will increase as time passes and then eventually die off when all the sugars have been consumed by the natural yeasts and bacteria.

Carbonation will mainly occur during your secondary fermentation. At this time you'll probably need to release some of the pressure built by the carbonation in the jar or bottle, which is called 'burping'. You can 'burp' your second ferment by gently opening the bottle/jar to release the pressure and then resealing it.

8. Do not touch!

Never touch your kombucha and jun scoby with metal objects, so remember to take off any jewellery and use wooden utensils.

9. Don't wash your scobys

You'll wash away all the good bacteria.

10. (Refrigerated) shelf life

All finished probiotic drinks can be stored in the refrigerator for 2–3 weeks.

RESTING SOLUTIONS

If you want to take a break from fermenting, there are a couple of resting solutions you can keep your cultures in while they (and you!) rest. This will keep them dormant until they are ready to use again. Here are some recipes for resting your cultures.

Resting water kefir solution

YOU NEED
50 g unrefined sugar
1 litre filtered or spring water
20 g kefir grains (activated)

Dissolve sugar in water and place in a jar along with kefir grains. Store in the refrigerator for up to 3 weeks.

Resting milk kefir

YOU NEED
2 teaspoons milk kefir grains
1 litre milk

Place kefir grains in a jar with milk and store in the refrigerator for up to 3 weeks.

Resting kombucha and jun

YOU NEED
1 scoby (see page 8)
120 ml starter tea (kombucha or jun)

Place scoby in starter tea and in the refrigerator. You can also start to build a scoby 'hotel' with your scoby babies in case your mother scoby dies or becomes contaminated.

If you have an overflow of scoby babies, gift them to friends and family. Or they can be composted.

Resting the Ginger Bug

You can rest your Ginger Bug (see page 108) in the refrigerator and feed it 2 tablespoons grated ginger and 2 tablespoons sugar every week.

TROUBLESHOOTING

What if your batch is too sour or too vinegar-y?

Your batch may have over-fermented. Try diluting kombucha, jun or wild fermentations with soda water or juice from fruit. If your kefir is over-fermented and has a funky smell, discard it and restart your fermentation again.

What if there's mould?

If you see mould, throw the batch and the culture away as it is not safe to drink.

What if there's no carbonation?

Once your batch of drink has been bottled, you may need to leave it longer to build carbonation, especially if the temperature in your house is too cold. It could take up to 5–7 days in colder weather and 1–2 days in warmer weather.

What if the kefir is slimy?

Your kefir may have over-fermented. Rinse your kefir grains and restart your fermentation again.

What if the milk kefir separates?

If milk kefir separates into curds and whey, you may have over-fermented it or the grain ratio may outweigh the milk. You'll need to remove some grains. Shake the jar of milk kefir so that the curds separate from the kefir grains. Pour the mixture over a sieve into a bowl and gently stir with a rubber spatula. You may need to add the whey that has been strained out back into the sieve (several times) to separate the grains from the curds.

What if the scoby doesn't float?

It's normal for a scoby to sink the first couple of days if it's too cold. If it doesn't float after a week it may not be a viable scoby. Source a new scoby and try again.

What if the wild fermented batch smells and tastes funky?

Don't drink it! If it smells rotten, your batch has probably gone bad.

KEFIRS

*Kefir comes from the Turkish word 'keif',
meaning good feeling. In this chapter
we explore both water and milk kefirs.*

Basic Water Kefir • Basic Coconut Water
Kefir • Basic Milk (or Coconut Milk) Kefir
Orange-sicle Water Kefir • Strawberry Ginger
Water Kefir • Pineapple Cucumber Water Kefir
Watermelon Mint Water Kefir • Pink Grapefruit
Water Kefir • Lime Lemongrass Coconut Water
Kefir • Passion Fruit Coconut Water Kefir
Cucumber Mint Coconut Water Kefir • Ginger
Turmeric Coconut Water Kefir • Breakfast Milk
Kefir Shake • Mint Cacao Milk Kefir Shake • Café
au Lait Milk Kefir Smoothie • Peanut Butter &
Cacao Milk Kefir • Piña Colada Coconut Milk
Kefir Shake • Mango Lassi Coconut Milk Kefir
Shake • Blueberry Coconut Milk Kefir Shake
Cashew Cinnamon Coconut Milk Kefir

BASIC WATER KEFIR

Makes: 1 litre
Preparation: 2–4 days

YOU NEED
1 litre spring water • 50 g unrefined sugar
5 g unactivated kefir grains (20 g activated) • 1 Medjool date, pitted

Combine water and sugar in a 1.5 litre jar; shake until sugar is dissolved.
Add remaining ingredients, cover with muslin and secure. Place jar out of direct
sunlight and ferment for 48 hours. Water will become cloudy, slightly effervescent
and sweet when ready. Strain kefir into a jar or bottle and use grains to start
another fermentation. Second fermentation: leave bottle at room temperature
with a tight-fitting lid for 1–2 days or until carbonated. 'Burp' kefir every day.

BASIC COCONUT WATER KEFIR

Makes: 1 litre
Preparation: 2–4 days

YOU NEED

1 litre coconut water • 5 g unactivated water kefir grains (20 g activated)

Combine water and grains in a 1.5 litre jar. Cover with a muslin and secure. Place jar out of direct sunlight and ferment for 48 hours. Water will become cloudy, slightly effervescent and sweet when ready. Strain kefir into a jar or bottle and transfer grains to start another fermentation. Second fermentation: leave bottle at room temperature with a tight-fitting lid for 1–2 days or until carbonated. 'Burp' kefir every day.

BASIC MILK (OR COCONUT MILK) KEFIR

Makes: 1 litre
Preparation: 1–2 days

YOU NEED

1 litre whole milk (or coconut milk) • 2 teaspoons activated milk kefir grains

Combine milk and grains in a 1.5 litre jar; stir together. Cover with a muslin and secure. Place jar out of direct sunlight but in a warm place. Ferment for 24–48 hours – milk will become tangy and thickened when ready. Strain grains into a bowl. Use a rubber spatula to gently push thicker milk kefir through sieve. Transfer grains to fresh milk to start another fermentation. Pour fermented milk kefir into a bottle.

ORANGE-SICLE WATER KEFIR

Makes: 1 litre
Preparation: 4–5 days

YOU NEED

1 batch of Basic Water Kefir (see page 16)

juice of 2 oranges • ½ teaspoon vanilla extract

V *Vitamin boosting*

Follow instructions for Basic Water Kefir. Once initial fermentation has finished,
remove grains and date from finished water kefir and set aside. Pour kefir into
a 1 litre bottle with a tight-fitting lid; add orange juice and vanilla. Tightly seal
bottle and ferment for 2–3 days in a cool, dry place. 'Burp' kefir every day.

STRAWBERRY GINGER WATER KEFIR

Makes: 1 litre
Preparation: 4–5 days

YOU NEED

1 batch of Basic Water Kefir (see page 16)

handful of strawberries (about 150 g), puréed or roughly chopped

3 cm piece of ginger, sliced

D *Aids digestion*

Follow instructions for Basic Water Kefir. Once initial fermentation has finished, scoop grains and date from finished water kefir and set aside. Pour kefir into a 1 litre bottle with a tight-fitting lid; add strawberry purée and ginger. Tightly seal bottle and ferment for 2–3 days in a cool, dry place. 'Burp' kefir every day.

PINEAPPLE CUCUMBER WATER KEFIR

Makes: 1 litre
Preparation: 4–5 days

YOU NEED
1 batch of Basic Water Kefir (see page 16)
100 g pineapple, puréed or roughly chopped • 1 cucumber, sliced

Follow instructions for Basic Water Kefir. Once initial fermentation has finished, scoop grains and date from finished water kefir and set aside. Pour kefir into a 1 litre bottle with a tight-fitting lid; add pineapple and cucumber. Tightly seal bottle and ferment for 2–3 days in a cool, dry place. 'Burp' kefir every day.

WATERMELON MINT WATER KEFIR

Makes: 1 litre
Preparation: 4–5 days

YOU NEED

1 batch of Basic Water Kefir (see page 16)

100 g watermelon, rind removed and thinly sliced • 4 mint sprigs

Follow instructions for Basic Water Kefir. Once initial fermentation has finished, scoop grains and date from finished water kefir and set aside. Pour kefir into a 1 litre bottle with a tight-fitting lid; add watermelon and mint. Tightly seal bottle and ferment for 2–3 days in a cool, dry place. 'Burp' kefir every day.

PINK GRAPEFRUIT WATER KEFIR

Makes: 1 litre
Preparation: 4–5 days

YOU NEED

1 batch of Basic Water Kefir (see page 16) • 1 pink grapefruit, juiced

Follow instructions for Basic Water Kefir. Once initial fermentation has finished,
scoop grains and date from finished water kefir and set aside. Pour kefir into
a 1 litre bottle with a tight-fitting cap; add grapefruit juice. Tightly seal bottle
and ferment for 2–3 days in a cool, dry place. 'Burp' kefir every day.

LIME LEMONGRASS COCONUT WATER KEFIR

Makes: 1 litre
Preparation: 4–5 days

YOU NEED
1 batch of Basic Coconut Water Kefir (see page 18)

1 lemongrass stalk, sliced • juice of 1 lime, plus 2 lime wedges or slices

I *Immune boosting*

Follow instructions for Basic Coconut Water Kefir. Once initial fermentation has finished, scoop grains from the finished water kefir and set aside. Pour kefir into a 1 litre bottle with a tight-fitting lid; add lemongrass and lime juice. Tightly seal bottle and ferment for 2–3 days in a cool, dry place. 'Burp' kefir every day. Garnish with lime slices or wedges.

PASSION FRUIT COCONUT WATER KEFIR

Makes: 1 litre
Preparation: 4–5 days

YOU NEED

1 batch of Basic Coconut Water Kefir (see page 18) • pulp from 2 passion fruits

Follow instructions for Basic Coconut Water Kefir. Once initial fermentation has finished, scoop grains from the finished water kefir and set aside. Pour kefir into a 1 litre bottle with a tight-fitting lid; add passion fruit pulp. Tightly seal bottle and ferment for 2–3 days in a cool, dry place. 'Burp' kefir every day.

CUCUMBER MINT COCONUT WATER KEFIR

Makes: 1 litre
Preparation: 4–5 days

YOU NEED
1 batch of Basic Coconut Water Kefir (see page 18)
1 cucumber, chopped • 4 mint sprigs

Follow instructions for Basic Coconut Water Kefir. Once initial fermentation has finished, scoop grains from finished water kefir and set aside. Pour kefir into a 1 litre bottle with a tight-fitting lid; add cucumber and mint. Tightly seal bottle and ferment for 2–3 days in a cool, dry place. 'Burp' kefir every day.

GINGER TURMERIC COCONUT WATER KEFIR

Makes: 1 litre
Preparation: 4–5 days

YOU NEED
1 batch of Basic Coconut Water Kefir (see page 18)
3 cm piece of ginger, thinly sliced
4 cm piece of fresh turmeric root, puréed (or 2 teaspoons ground turmeric)

Follow instructions for Basic Coconut Water Kefir. Once initial fermentation has
finished, scoop grains from finished water kefir and set aside. Pour kefir into a
1 litre bottle with a tight-fitting cap; add ginger and turmeric. Tightly seal bottle
and ferment for 2–3 days in a cool, dry place. 'Burp' kefir every day.

BREAKFAST MILK KEFIR SHAKE

Makes: 250 ml
Preparation: 3 minutes

YOU NEED

240 ml Basic Milk Kefir (see page 20) • ½ frozen banana, peeled
4 Medjool dates, pitted • 1 teaspoon maca powder
ground cinnamon, to garnish

Combine all ingredients, except for cinnamon, in a blender and blend until smooth. Pour into a glass and serve immediately. Garnish with cinnamon.

MINT CACAO MILK KEFIR SHAKE

Makes: 250 ml
Preparation: 3–5 minutes

YOU NEED
100 ml Basic Coconut Milk Kefir (see page 20) • handful of mint leaves

½ teaspoon spirulina • ½ teaspoon vanilla extract

1 Medjool date, pitted • 1 tablespoon raw cacao nibs

Combine all ingredients in a blender and blend until smooth.
Pour into a glass and serve.

CAFÉ AU LAIT MILK KEFIR SMOOTHIE

Makes: 250 ml
Preparation: 3 minutes

YOU NEED
240 ml Basic Milk Kefir (see page 20) • 2 teaspoons instant coffee
1 teaspoon agave syrup • 220 g ice • cacao nibs, to serve

Antioxidant

Combine all ingredients, except cacao nibs, in a blender and blend until smooth.
Pour into a glass and top with cacao nibs. Serve.

PEANUT BUTTER & CACAO MILK KEFIR

Makes: 250 ml
Preparation: 3 minutes

YOU NEED
240 ml Basic Milk Kefir (see page 20) • 120 ml plain, low-fat yoghurt
2 tablespoons peanut butter • 2 tablespoons cacao nibs
1 Medjool date, pitted • 220 g ice

A *Antioxidant*

Combine all ingredients in a blender and blend until smooth.
Pour into a glass and serve.

PIÑA COLADA COCONUT MILK KEFIR SHAKE

Makes: 250 ml
Preparation: 3 minutes

YOU NEED

240 ml Basic Coconut Milk Kefir (see page 20)

70 g frozen pineapple, chopped • ½ frozen banana • 1 teaspoon honey

1 *Anti-inflammatory*

Combine all ingredients in a blender and blend until smooth.
Pour into a glass and serve.

MANGO LASSI COCONUT MILK KEFIR SHAKE

Makes: 250 ml
Preparation: 3 minutes

YOU NEED
240 ml Basic Coconut Milk Kefir (see page 20) • 120 ml coconut yoghurt
75 g frozen mango, roughly chopped • ½ teaspoon ground turmeric
1 tablespoon honey • pinch of salt

c *Prevents cancer*

Combine all ingredients in a blender and blend until smooth.
Pour into a glass and serve.

BLUEBERRY COCONUT MILK KEFIR SHAKE

Makes: 250 ml
Preparation: 3 minutes

YOU NEED
240 ml Basic Coconut Milk Kefir (see page 20) • 140 g frozen blueberries
1 tablespoon honey • 20 g grated coconut • 1 teaspoon bee pollen

A *Antioxidant*

Combine all ingredients, except bee pollen, in a blender and blend until smooth.
Pour into a glass and top with bee pollen. Serve.

CASHEW CINNAMON COCONUT MILK KEFIR

Makes: 250 ml
Preparation: 3 minutes

YOU NEED
240 ml Basic Coconut Milk Kefir (see page 20) • 120 ml coconut yoghurt
60 g raw cashews • ½ teaspoon vanilla extract • 1 tablespoon honey
pinch of salt • ½ teaspoon ground cinnamon, plus extra to garnish

B *Anti-bloating*

Combine all ingredients in a blender and blend until smooth.
Pour into a glass and serve sprinkled with cinnamon.

KOMBUCHA

*Kombucha has been called the
'Immortal Health Elixir' by the Chinese
for its versatile health benefits, but most
importantly, it's tasty!*

Basic Kombucha • Hibiscus & Lime Kombucha
Lemon Verbena Oolong Kombucha
Apple Cider Kombucha • Winter Citrus
Kombucha • Earl Grey & Lavender Kombucha
Watermelon Jalapeño Kombucha • Carrot
Orange Kombucha • Pear & Spice Kombucha
Herbal Kombucha • Plum Darjeeling Kombucha

BASIC KOMBUCHA

Makes: 1.5 litres
Preparation: 4–10 days

YOU NEED

2 litres spring water • 3 black tea bags (or 1 tablespoon tea leaves)
6 tablespoons raw sugar • 1 kombucha scoby (see page 10)
120 ml kombucha starter liquid (from previous kombucha or see page 10)

D Detoxifier

Bring half the water to a simmer in a pan. Add tea and steep for 3–5 minutes. Strain into a jar and add sugar; stir to dissolve. Add remaining water and cool. Add remaining ingredients, cover with muslin and secure. Place out of direct sunlight and ferment for 4–5 days. Remove scoby and 120 ml kombucha and set aside to reuse or rest. Pour remaining kombucha into a 2 litre bottle with tight-fitting lid. Second fermentation: leave at room temperature for 4–5 days to build carbonation. 'Burp' every 2 days.

HIBISCUS & LIME KOMBUCHA

Makes: 1.5 litres
Preparation: 4–10 days

YOU NEED

1 batch of Basic Kombucha (see page 58), replacing black tea with hibiscus tea

juice from 1 lime, plus a lime wedge to garnish

Follow instructions for Basic Kombucha. Once initial fermentation has finished, remove scoby and 120 ml kombucha and set aside to reuse or rest. Pour kombucha into a 2 litre bottle with a tight-fitting lid and add lime juice. Leave at room temperature for 4–5 days to build carbonation. 'Burp' every 2 days. Garnish with a lime wedge to serve.

LEMON VERBENA OOLONG KOMBUCHA

Makes: 1.5 litres
Preparation: 4–10 days

YOU NEED
1 batch of Basic Kombucha (see page 58), replacing black tea with oolong tea
30 g lemon verbena, roughly chopped

Follow instructions for Basic Kombucha. Once initial fermentation has finished, remove scoby and 500 ml kombucha and set aside to reuse or rest. Second fermentation: pour kombucha into a 2 litre bottle with a tight-fitting lid and add lemon verbena. Leave at room temperature for 4–5 days to build carbonation. 'Burp' every 2 days. Strain out lemon verbena and bottle into a 2 litre bottle.

APPLE CIDER KOMBUCHA

Makes: 1.5 litres
Preparation: 4–10 days

YOU NEED

1 batch of Basic Kombucha (see page 58)

2 apples (find an apple variety you like), finely chopped

4 cloves • 2 cinnamon sticks • 1 star anise

Follow instructions for Basic Kombucha. Once initial fermentation has finished, remove scoby and 500 ml kombucha and set aside to reuse or rest. Second fermentation: pour kombucha into a 2 litre bottle with a tight-fitting lid and add apples and spices. Leave at room temperature for 4–5 days to build carbonation. 'Burp' every 2 days. Strain and bottle into a 2 litre bottle.

WINTER CITRUS KOMBUCHA

Makes: 1.5 litres
Preparation: 4–10 days

YOU NEED
1 batch of Basic Kombucha (see page 58), replacing black tea with English Breakfast
juice of 1 blood orange • juice of ½ grapefruit • juice of 2 lemons

Follow instructions for Basic Kombucha. Once initial fermentation has finished, remove scoby and 500 ml kombucha and set aside to reuse or rest. Second fermentation: pour kombucha into a 2 litre bottle with a tight-fitting lid and add citrus juices. Leave at room temperature for 4–5 days to build carbonation. 'Burp' every 2 days. Strain and bottle into a 2 litre bottle.

EARL GREY & LAVENDER KOMBUCHA

Makes: 1.5 litres
Preparation: 4–10 days

YOU NEED
1 batch of Basic Kombucha (see page 58), replacing black tea with Earl Grey tea
2 tablespoons lavender buds

A *Antioxidant*

Follow instructions for Basic Kombucha. Once initial fermentation has finished, remove scoby and 500 ml kombucha and set aside to reuse or rest. Second fermentation: pour kombucha into a 2 litre bottle with a tight-fitting lid and add lavender. Leave at room temperature for 4–5 days to build carbonation. 'Burp' every 2 days. Strain and bottle into a 2 litre bottle.

WATERMELON JALAPEÑO KOMBUCHA

Makes: 1.5 litres
Preparation: 4–10 days

YOU NEED

1 batch of Basic Kombucha (see page 58), replacing black tea with white tea

150 g watermelon, finely chopped • ½ jalapeño, finely chopped

H *Hydrating*

Follow instructions for Basic Kombucha. Once initial fermentation has finished, remove scoby and 500 ml kombucha and set aside to reuse or rest. Second fermentation: pour kombucha into a 2 litre bottle with a tight-fitting lid and add watermelon and jalapeño. Leave at room temperature for 4–5 days to build carbonation. 'Burp' every 2 days. Strain and bottle into a 2 litre bottle.

71

CARROT ORANGE KOMBUCHA

Makes: 1.5 litres
Preparation: 4–10 days

YOU NEED
1 batch of Basic Kombucha (see page 58) • 120 ml carrot juice
120 ml orange juice

v *Vitamin boosting*

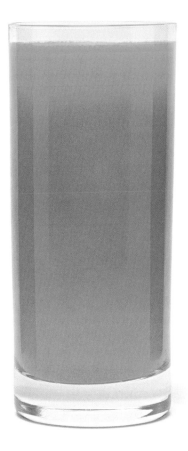

Follow instructions for Basic Kombucha. Once initial fermentation has finished, remove scoby and 500 ml kombucha and set aside to reuse or rest. Second fermentation: pour into a 2 litre bottle with a tight-fitting lid and add carrot juice and orange juice. Leave at room temperature for 4–5 days to build carbonation. 'Burp' every 2 days. Strain and bottle into a 2 litre bottle.

PEAR & SPICE KOMBUCHA

Makes: 1.5 litres
Preparation: 4–10 days

YOU NEED
1 batch of Basic Kombucha (see page 58), replacing black tea with white tea
120 ml pear juice • 4–6 cloves • 1 cinnamon stick • 2 star anise

Follow instructions for Basic Kombucha. Once initial fermentation has finished, remove scoby and 500 ml kombucha and set aside to reuse or rest. Second fermentation: pour kombucha into a 2 litre bottle with a tight-fitting lid and add pear juice and spices. Leave at room temperature for 4–5 days to build carbonation. 'Burp' every 2 days. Strain and bottle into a 2 litre bottle.

HERBAL KOMBUCHA

Makes: 1.5 litres
Preparation: 4–10 days

YOU NEED

1 batch of Basic Kombucha (see page 58), replacing black tea with
yerba mate or lapsang souchong

4 tablespoons hibiscus leaves • 4 holy basil sprigs

4 mint sprigs • 4 cm piece of ginger, chopped

Ⓐ *Antioxidant*

Follow instructions for Basic Kombucha, but brew tea with hibiscus leaves. Once initial fermentation has finished, remove scoby and 500 ml kombucha and set aside to reuse or rest and strain and discard the hibiscus. Second fermentation: pour kombucha into a 2 litre bottle with a tight-fitting lid and add remaining ingredients. Leave at room temperature for 4–5 days to build carbonation. 'Burp' every 2 days. Strain and bottle into a 2 litre bottle.

PLUM DARJEELING KOMBUCHA

Makes: 1.5 litres
Preparation: 4–10 days

YOU NEED

1 batch of Basic Kombucha (see page 58), replacing black tea with Darjeeling
100 g plums, stoned and finely chopped

G *Prevents cancer*

Follow instructions for Basic Kombucha. Once initial fermentation has finished, remove scoby and 500 ml kombucha and set aside to reuse or rest. Second fermentation: pour kombucha into a 2 litre bottle with a tight-fitting lid and add plums. Leave at room temperature for 4–5 days to build carbonation. 'Burp' every 2 days. Strain and bottle into a 2 litre bottle.

JUN

*Jun is cousin to kombucha but rather
than using black tea and sugar,
it uses green or white tea and honey.*

Basic Jun • Jun with Jasmine Tea & Peach • Jun with
Mixed Berries • Jun with Pink Lady Apples
Jun with Strawberries & Thyme • Jun with
Lavender & Blueberries • White Jun with
Elderflowers • Jun with Blackberry & Lemon Balm
Jun with Grapefruit & Cardamom • Jun with
Blueberry & Vanilla

BASIC JUN

Makes: 1 litre
Preparation: 4–10 days

YOU NEED
1 litre spring water • 2 green tea teabags or 1 teaspoon loose-leaf green tea
4 tablespoons raw honey • 1 jun scoby (see page 10)
60 ml jun starter liquid (from previous jun or see page 10)

Bring water to a simmer in a pan. Add tea and steep for 3–5 minutes. Strain into a jar and add honey; stir to dissolve, then cool. Add scoby and starter liquid, cover and secure. Place jar out of direct sunlight and ferment for 4–5 days or longer in cooler weather. Remove scoby and 120 ml jun and set aside to reuse or rest. Pour remaining jun into a bottle with tight-fitting lid. Second fermentation: leave at room temperature for 4–5 days to build carbonation. 'Burp' every 2 days.

JUN WITH JASMINE TEA & PEACH

Makes: 1 litre
Preparation: 4–10 days

YOU NEED

1 batch of Basic Jun (see page 82), replacing green tea with jasmine tea

2 peaches, puréed

Follow instructions for Basic Jun. Once initial fermentation has finished, remove scoby and 500 ml jun and set aside to reuse or rest. Pour jun into a 1 litre bottle with a tight-fitting lid and add peach purée. Leave at room temperature for 4–5 days to build carbonation. 'Burp' every 2 days. Strain into a 1 litre bottle.

JUN WITH MIXED BERRIES

Makes: 1 litre
Preparation: 4–10 days

YOU NEED
1 batch of Basic Jun (see page 82), replacing green tea with jasmine tea
100 g mixed berries, puréed

(A) *Antioxidant*

Follow instructions for Basic Jun. Once initial fermentation has finished, remove scoby and 500 ml jun and set aside to reuse or rest. Pour jun into a 1 litre bottle with a tight-fitting lid and add berry purée. Leave at room temperature for 4–5 days to build carbonation. 'Burp' every 2 days. Strain into a 1 litre bottle.

JUN WITH PINK LADY APPLES

Makes: 1 litre
Preparation: 4–10 days

YOU NEED

1 batch of Basic Jun (see page 82) • 2 Pink Lady apples, peeled, cored and puréed

(A) *Antioxidant*

Follow instructions for Basic Jun. Once initial fermentation has finished, remove scoby and 500 ml jun and set aside to reuse or rest. Pour jun into a 1 litre bottle with a tight-fitting lid and add apple purée. Leave at room temperature for 4–5 days to build carbonation. 'Burp' every 2 days. Strain into a 1 litre bottle.

JUN WITH STRAWBERRIES & THYME

Makes: 1 litre
Preparation: 4–10 days

YOU NEED

1 batch of Basic Jun (see page 82), replacing green tea with white tea

200 g strawberries, puréed • 6 thyme sprigs

Follow instructions for Basic Jun. Once initial fermentation has finished, remove
scoby and 500 ml jun and set aside to reuse or rest. Pour jun into a 1 litre bottle with
a tight-fitting lid and add strawberry purée and thyme. Leave at room temperature
for 4–5 days to build carbonation. 'Burp' every 2 days. Strain into a 1 litre bottle.

JUN WITH LAVENDER & BLUEBERRIES

Makes: 1 litre
Preparation: 4–10 days

YOU NEED

1 batch of Basic Jun (see page 82), replacing green tea with white tea

150 g blueberries, puréed • 4 tablespoons lavender buds

Follow instructions for Basic Jun. Once initial fermentation has finished, remove
scoby and 500 ml jun and set aside to reuse or rest. Pour jun into a 1 litre bottle with
a tight-fitting lid and add blueberry purée and lavender. Leave at room temperature
for 4–5 days to build carbonation. 'Burp' every 2 days. Strain into a 1 litre bottle.

WHITE JUN WITH ELDERFLOWERS

Makes: 1 litre
Preparation: 4–10 days

YOU NEED
1 batch of Basic Jun (see page 82), replacing green tea with white tea
4 tablespoons dried elderflowers

I *Anti-inflammatory*

Follow instructions for Basic Jun. Once initial fermentation has finished, remove scoby and 500 ml jun and set aside to reuse or rest. Pour jun into a 1 litre bottle with a tight-fitting lid and add elderflowers. Leave at room temperature for 4–5 days to build carbonation. 'Burp' every 2 days. Strain into a 1 litre bottle.

JUN WITH BLACKBERRY & LEMON BALM

Makes: 1 litre
Preparation: 4–10 days

YOU NEED

1 batch of Basic Jun (see page 82), replacing green tea with white tea

125 g blackberries, puréed • 6 sprigs lemon balm

ⓥ *Vitamin boosting*

Follow instructions for Basic Jun. Once initial fermentation has finished,
remove jun scoby and 500 ml jun and set aside to reuse or rest. Pour jun into
a 1 litre bottle with a tight-fitting lid and add blackberry purée and lemon balm.
Leave at room temperature for 4–5 days to build carbonation. 'Burp' every 2 days.
Strain into a 1 litre bottle.

JUN WITH GRAPEFRUIT & CARDAMOM

Makes: 1 litre
Preparation: 4–10 days

YOU NEED

1 batch of Basic Jun (see page 82), replacing green tea with white tea

juice of 1 pink grapefruit • 3 cardamom pods, lightly smashed

Immune boosting

Follow instructions for Basic Jun. Once initial fermentation has finished, remove jun scoby and 500 ml jun and set aside to reuse or rest. Pour jun into a 1 litre bottle with a tight-fitting lid and add grapefruit juice and cardamom. Leave at room temperature for 4–5 days to build carbonation. 'Burp' every 2 days. Strain into a 1 litre bottle.

JUN WITH BLUEBERRY & VANILLA

Makes: 1 litre
Preparation: 4–10 days

YOU NEED

1 batch of Basic Jun (see page 82), replacing green tea with white tea

150 g blueberries, puréed • 1 vanilla pod, split lengthways

(A) *Antioxidant*

Follow instructions for Basic Jun. Once initial fermentation has finished, remove
scoby and 500 ml jun and set aside to reuse or rest. Pour jun into a 1 litre
bottle with a tight-fitting lid and add blueberry purée and vanilla pod.
Leave at room temperature for 4–5 days to build carbonation.
'Burp' every 2 days. Strain into a 1 litre bottle.

NATURE'S FERMENTATION

Here you'll find fermentations that only require fruit or vegetables and a sweetener.

Basic Beetroot Kvass • Basic Pineapple Tepache
Basic Ginger Bug • Spiced Beetroot Kvass
Orange Ginger Beetroot Kvass • Pomegranate
Beetroot Kvass • Blackberry Vanilla Tepache
Pineapple Black Pepper Tepache • Lemon &
Lavender Pineapple Tepache • Ginger Beer
Ginger Lime Mint Soda • Pineapple Ginger Soda
Persimmon & Vanilla Ginger Soda • Strawberry
Shrub • Concord Grape Shrub • Pear Cinnamon
Shrub • Peach Champagne Shrub • Tangerine
Rice Vinegar Shrub

BASIC BEETROOT KVASS

Makes: about 700 ml
Preparation: 3–8 days

YOU NEED

340 g beetroot, washed, diced and cut into 2 cm cubes

1 teaspoon sea salt (uniodised) • 700 ml spring or filtered water

Place beetroot in a 1 litre jar with a tight-fitting lid and add salt. Pour water into jar, leaving 2 cm head space between water and lid. Leave jar on a plate (for any spillage) at room temperature for 3–5 days, stirring or shaking daily. Taste kvass to see if it's ready – it should be a little salty and sour. Strain into a bottle. To add carbonation, tightly seal and leave at room temperature for 2–3 days.

BASIC PINEAPPLE TEPACHE

Makes: 2 litres
Preparation: 3–8 days

YOU NEED

½ pineapple (about 1 kg), unpeeled, top removed and cut into 2 cm pieces

220 g light brown sugar • 6 cm piece of ginger, roughly chopped

2 litres spring or filtered water

Place pineapple in a large jar, add sugar and 240 ml of the water. Close lid and shake to dissolve sugar. Add ginger and remaining water and stir. Cover jar with double layer of muslin and secure with a rubber band. Place in a dry place away from direct sunlight and ferment for 3–5 days, stirring daily. Strain and discard pineapple and ginger. Pour liquid into a 2 litre bottle with a tight-fitting lid, leaving about 2 cm head space. Leave at room temperature for 1–3 days to carbonate. 'Burp' daily.

BASIC GINGER BUG

Makes: 500 ml
Preparation: 4–6 days

YOU NEED

2 tablespoons grated ginger, plus extra for feeding

2 tablespoons raw sugar, plus extra for feeding

750 ml spring water

D *Aids digestion*

Place ginger and sugar in a jar with 500 ml of the water; stir. Cover, secure and place at room temperature. Every other day, add (feed with) 2 tablespoons grated ginger and 2 tablespoons sugar. Keep feeding for 4–6 days until fermented – ginger will float to top and you'll see bubbles. Strain and retain half the liquid to keep an active bug going; discard ginger bits and use for other drinks. Add remaining water and 1 teaspoon grated ginger and sugar back into new batch to keep it alive. Add 1 teaspoon ginger and 1 teaspoon sugar once a week and stir.

SPICED BEETROOT KVASS

Makes: about 700 ml
Preparation: 3–8 days

YOU NEED
1 batch of Basic Beetroot Kvass (see page 104) • 1–2 cinnamon sticks
2–3 star anise • 4 cm piece of ginger, chopped

B *Anti-bloating*

Follow instructions for Basic Beetroot Kvass, adding 1 cinnamon stick, 2 star anise
and ginger. Strain fermented kvass liquid into a 1 litre bottle. Add remaining
cinnamon stick and star anise if needed. Add carbonation by tightly sealing bottle
and leaving in a cool, dry place for 2–3 days. Refrigerate and serve chilled.

ORANGE GINGER BEETROOT KVASS

Makes: about 700 ml
Preparation: 3–8 days

YOU NEED

1 batch of Basic Beetroot Kvass (see page 104)
2 wide strips of orange peel (pith removed)
juice of 1 orange • 4 cm piece of ginger, chopped

v *Vitamin boosting*

Follow instructions for Basic Beetroot Kvass, adding remaining ingredients. Strain fermented kvass liquid into a 1 litre bottle. Add carbonation by tightly sealing bottle and leaving in a cool, dry place for 2–3 days. Refrigerate and serve chilled.

POMEGRANATE BEETROOT KVASS

Makes: 1 litre
Preparation: 3–8 days

YOU NEED

1 batch of Basic Beetroot Kvass (see page 104) • 250 ml pomegranate juice

Antioxidant

Follow instructions for Basic Beetroot Kvass and add pomegranate juice.
Strain fermented kvass liquid into a 1 litre bottle. Add carbonation by tightly sealing
bottle and leaving in a cool, dry place for 2–3 days. Refrigerate and serve chilled.

BLACKBERRY VANILLA TEPACHE

Makes: 1.25 litres
Preparation: 3–8 days

YOU NEED

1 batch of Basic Pineapple Tepache (see page 106)

100 g blackberries • 1 vanilla pod, split lengthways

Antioxidant

Follow instructions for Basic Pineapple Tepache. Strain and discard pineapple and ginger, add blackberries and vanilla and tightly seal with a lid, leaving 2 cm head space. Leave at room temperature for 1–3 days. 'Burp' daily to release pressure. Strain and pour liquid into a 2 litre bottle with a tight-fitting lid. Refrigerate and serve cold.

PINEAPPLE BLACK PEPPER TEPACHE

Makes: 2 litres
Preparation: 3–8 days

YOU NEED
1 batch of Basic Pineapple Tepache (see page 106) • 6 black peppercorns

Follow instructions for Basic Pineapple Tepache, adding peppercorns. Strain and
discard pineapple, ginger and peppercorns. Tightly seal with a lid, leaving
2 cm head space. Leave at room temperature for 1–3 days. 'Burp' daily to release
pressure. Strain and pour liquid into a 2 litre bottle with a tight-fitting lid.
Refrigerate and serve cold.

LEMON & LAVENDER PINEAPPLE TEPACHE

Makes: 2 litres
Preparation: 3–8 days

YOU NEED

1 batch of Basic Pineapple Tepache (see page 106) • peel of 1 lemon, pith removed

2 teaspoons lavender buds

Follow instructions for Basic Pineapple Tepache. Strain and discard pineapple
and ginger. Tightly seal with a lid, leaving 2 cm head space. Add lemon peel
and lavender. Leave at room temperature for 1–3 days. 'Burp' daily to release
pressure. Strain and pour liquid into a 2 litre bottle with a tight-fitting lid.
Refrigerate and serve cold.

GINGER BEER

Makes: 1 litre
Preparation: 4–8 days

YOU NEED
12 cm piece of ginger, grated • 1 litre filtered or spring water • 100 g raw sugar
juice of 2 lemons • 125 ml Basic Ginger Bug (see page 108)

Place grated ginger and 500 ml of the water in a saucepan. Cover and simmer for 15 minutes, then add sugar; stir until dissolved. Leave to cool and add lemon juice and Ginger Bug. Add remaining water. Pour into a 1 litre bottle with a tight-fitting lid and seal. Leave at room temperature for 1–2 days. This tends to carbonate quickly. 'Burp' daily to release pressure. Serve chilled.

GINGER LIME MINT SODA

Makes: 1 litre
Preparation: 4–8 days

YOU NEED

2 tablespoons grated ginger • 1 litre spring water • 100 g raw sugar
juice of 3–4 limes • 125 ml Basic Ginger Bug (see page 108) • 25 g mint leaves

Place grated ginger and 500 ml of the water in a saucepan. Cover and simmer for
15 minutes, then add sugar; stir until dissolved. Leave liquid to cool and add lime
juice and Ginger Bug. Add remaining water. Pour into a 1 litre bottle with
a tight-fitting lid, add mint and seal. Leave at room temperature for 1–2 days.
This tends to carbonate quickly. 'Burp' daily to release pressure. Serve chilled.

PINEAPPLE GINGER SODA

Makes: 1 litre
Preparation: 4–8 days

YOU NEED
4 tablespoons grated ginger • 1 litre spring water • 100 g raw sugar
250 g pineapple, chopped • 125 ml Basic Ginger Bug (see page 108)

Place grated ginger and 500 ml of the water in a saucepan. Cover and simmer for
15 minutes, then add sugar; stir until dissolved. Leave to cool and add pineapple
and Ginger Bug. Muddle pineapple and add remaining water. Pour into a 1 litre
bottle with a tight-fitting lid and seal. Leave at room temperature for 1–2 days.
This tends to carbonate quickly. 'Burp' daily to release pressure. Serve chilled.

PERSIMMON & VANILLA GINGER SODA

Makes: 1 litre
Preparation: 4–8 days

YOU NEED

4 tablespoons grated ginger • 1 litre spring water • 100 g raw sugar
3 persimmons, puréed • 1 vanilla pod, split lengthways
125 ml Basic Ginger Bug (see page 108)

A *Antioxidant*

Place grated ginger and 500 ml of the water in a saucepan. Cover and simmer for
15 minutes, then add sugar; stir until dissolved. Leave to cool and add persimmon
purée, vanilla pod and Ginger Bug. Add remaining water. Pour into a 1 litre bottle
with a tight-fitting lid and seal. Leave at room temperature for 1–2 days. This tends
to carbonate quickly. 'Burp' daily to release pressure. Serve chilled.

STRAWBERRY SHRUB

Makes: 1 litre
Preparation: 1–2 days

YOU NEED
450 g strawberries, quartered • 200 g unrefined, granulated sugar
470 ml red wine vinegar

Combine strawberries with sugar and stir together until fruit is covered with sugar. Chill for 1–2 days until fruit is macerated. Strain syrup, pressing the strawberries to push out any extra juices. Add vinegar and whisk to combine. Pour shrub into a clean 1 litre bottle and refrigerate for up to 2 weeks. To serve, pour 30 ml shrub syrup into a glass filled with ice and top with soda water.

CONCORD GRAPE SHRUB

Makes: 1 litre
Preparation: 1–2 days

YOU NEED

450 g concord grapes, rinsed • 200 g unrefined, granulated sugar

470 ml white wine vinegar

Combine grapes with sugar and slightly muddle together until fruit is covered with
sugar. Chill for 1–2 days until fruit is macerated. Strain syrup, pressing the grapes to
push out any extra juices. Add vinegar and whisk to combine. Pour shrub into a clean
1 litre bottle and refrigerate for up to 2 weeks. To serve, pour 30 ml shrub syrup into
a glass filled with ice and top with soda water.

PEAR CINNAMON SHRUB

Makes: 1 litre
Preparation: 1–2 days

YOU NEED

450 g pears, peeled and chopped • 1 cinnamon stick, plus extra to serve
380 g light brown sugar • 470 ml white wine vinegar

Combine chopped pears and cinnamon stick with sugar and stir together until pears are covered with sugar. Chill for 1–2 days until fruit is macerated. Strain syrup, pressing the fruit to push out any extra juices. Add vinegar and whisk to combine. Pour shrub into a clean 1 litre bottle and refrigerate for up to 2 weeks. To serve, pour 30 ml shrub syrup into a glass filled with ice and top with soda water.

PEACH CHAMPAGNE SHRUB

Makes: 1 litre
Preparation: 1–2 days

YOU NEED

450 g peaches, stoned and halved • 200 g unrefined, granulated sugar

470 ml Champagne vinegar

Combine peaches with sugar and stir together until fruit is covered with sugar. Chill for 1–2 days until fruit is macerated. Strain syrup, pressing the fruit to push out any extra juices. Add vinegar and whisk to combine. Pour shrub into a clean 1 litre bottle and refrigerate for up to 2 weeks. To serve, pour 30 ml shrub syrup into a glass filled with ice and top with soda water.

TANGERINE RICE VINEGAR SHRUB

Makes: 1 litre
Preparation: 1–2 days

YOU NEED
240 ml tangerine juice (about 5–6 tangerines) • peel from 4 tangerines
200 g unrefined, granulated sugar • 470 ml rice vinegar

Combine tangerine juice and peels with sugar and stir together until peels are covered with sugar. Chill for 1–2 days until fruit is macerated. Strain syrup, pressing the tangerine peels to push out any extra juices. Add vinegar and whisk to combine. Pour shrub into a clean 1 litre bottle and refrigerate for up to 2 weeks. To serve, pour 30 ml shrub syrup into a glass filled with ice and top with soda water.

WHEY-FERMENTED SODA

Whey is a by-product of yoghurt and kefir and can be used to create a different kind of probiotic drink. This chapter will guide you through the process of obtaining whey and what to do with it.

Basic Whey • Whey Lacto-fermented
Root Beer • Whey Lacto-fermented Lemon
Soda • Lacto-fermented Orange Vanilla Soda
Lacto-fermented Rosewater • Lacto-fermented
Summer Melon Soda • Lacto-fermented Grape
& Mint Soda • Lacto-fermented Plum
& Thyme Soda

BASIC WHEY

Makes: 1.5 litres
Preparation: 4–6 hours

YOU NEED
485 g plain, full-fat yoghurt or milk kefir (see page 20)

Line a sieve with two layers of a muslin or a clean tea towel and set sieve over a large bowl. Spoon yoghurt or milk kefir onto a muslin. Cover and leave at room temperature for 4–6 hours or refrigerate overnight. The liquid whey will drain into the bowl. Whey can be kept in the fridge for up to 2 weeks.

WHEY LACTO-FERMENTED ROOT BEER

Makes: 1.5 litres
Preparation: 3–6 days

YOU NEED

1 litre spring water • 1 tablespoon sarsaparilla bark • 1 tablespoon sassafras bark
3 star anise • 1 vanilla pod, split lengthways and seeds scraped
120 ml molasses • 120 g light brown sugar • 240 ml Basic Whey (see page 142)

A *Antioxidant*

Combine all ingredients, except Basic Whey, in a small saucepan and simmer for 20–30 minutes, covered, then cool. Strain and pour into a large jar; add whey. Cover jar with double layer of a muslin and secure with rubber band. Place in a dry place away from direct sunlight to ferment for 2–3 days. Transfer mixture to a bottle with a tight-fitting lid. Leave at room temperature for 1–3 days to carbonate. 'Burp' daily.

WHEY LACTO-FERMENTED LEMON SODA

Makes: 1.5 litres
Preparation: 3–6 days

YOU NEED
1 litre spring water • 2 tablespoons dried elderflowers • 120–240 ml honey
240 ml freshly squeezed lemon juice • 120 ml Basic Whey (see page 142)

Combine all ingredients, except Basic Whey, in a small saucepan and simmer for 20–30 minutes, covered, then cool. Strain and pour into a large jar; add whey. Cover jar with double layer of a muslin and secure with rubber band. Place in a dry place away from direct sunlight to ferment for 2–3 days. Transfer mixture to a bottle with a tight-fitting lid. Leave at room temperature for 1–3 days to carbonate. 'Burp' daily.

LACTO-FERMENTED ORANGE VANILLA SODA

Makes: 1.5 litres
Preparation: 3–6 days

YOU NEED
1 litre spring water • 120–240 ml honey
240 ml freshly squeezed orange juice • 2 thin peels of orange
120 ml Basic Whey (see page 142) • 1 vanilla pod, split lengthways and seeds scraped

Combine all ingredients, except Basic Whey, orange peels and vanilla, in a small saucepan and simmer for 20–30 minutes, covered, then cool. Strain and pour into a large jar; add whey. Cover jar with double layer of a muslin and secure with rubber band. Place in a dry place away from direct sunlight to ferment for 2–3 days. Transfer mixture to a bottle with a tight-fitting lid; add vanilla and orange peels to bottle. Leave at room temperature for 1–3 days to carbonate. 'Burp' daily.

LACTO-FERMENTED ROSEWATER

Makes: 1.5 litres
Preparation: 3–6 days

YOU NEED
1 litre spring water • 120–240 ml honey • 240 ml Basic Whey (see page 142)
1 tablespoon edible rose petals, plus extra to garnish

Combine all ingredients, except Basic Whey and rose petals, in a small saucepan and simmer for 20–30 minutes, covered, then cool. Strain and pour into a large jar; add whey. Cover jar with double layer of a muslin and secure with rubber band. Place in a dry place away from direct sunlight to ferment for 2–3 days. Transfer mixture to a bottle with a tight-fitting lid; add rose petals to bottle. Leave at room temperature for 1–3 days to carbonate. 'Burp' daily.

LACTO-FERMENTED SUMMER MELON SODA

Makes: 1.5 litres
Preparation: 3–6 days

YOU NEED

650 g mixed melon flesh, chopped and puréed • 1 litre spring water
120–240 ml honey • 240 ml Basic Whey (see page 142)

Strain melon purée through a muslin and discard the flesh. Combine all
ingredients, except Basic Whey, in a small saucepan. Add melon juice and simmer
for 20–30 minutes, covered, then cool. Strain and pour into a large jar; add whey.
Cover jar with double layer of a muslin and secure with rubber band. Place in a dry
place away from direct sunlight to ferment for 2–3 days. Transfer mixture to a bottle
with a tight-fitting lid. Leave at room temperature for 1–3 days to carbonate.
'Burp' daily.

LACTO-FERMENTED GRAPE & MINT SODA

Makes: 1.5 litres
Preparation: 3–6 days

YOU NEED

1 litre spring water • 225 g concord grapes • 120–240 ml honey
240 ml Basic Whey (see page 142) • 6 mint sprigs

Combine water, grapes and honey in a small saucepan and simmer for
20–30 minutes, covered, then cool. Strain and pour into a large jar; add whey.
Cover jar with double layer of a muslin and secure with rubber band. Place in
a dry place away from direct sunlight to ferment for 2–3 days. Transfer mixture
to a bottle with a tight-fitting lid and add mint. Leave at room temperature
for 1–3 days to carbonate. 'Burp' daily.

LACTO-FERMENTED PLUM & THYME SODA

Makes: 1.5 litres
Preparation: 3–6 days

YOU NEED

1 litre spring water • 4 plums, stoned • 120–240 ml honey
240 ml Basic Whey (see page 142) • 6 thyme sprigs

Combine water, plums and honey in a small saucepan and simmer for 20–30 minutes, covered, then cool. Strain and pour into a large jar; add whey. Cover jar with double layer of a muslin and secure with rubber band. Place in a dry place away from direct sunlight to ferment for 2–3 days. Transfer mixture to a bottle with a tight-fitting lid and add thyme. Leave at room temperature for 1–3 days to carbonate. 'Burp' daily.

INDEX

Entries in italics refer to recipe names.

A
agave syrup 44
Apple Cider Kombucha 64
apples 64, 88

B
banana 40, 48
Basic Beetroot Kvass 104, 110, 112, 114
Basic Coconut Milk Kefir 20, 42, 48, 50, 52, 54
Basic Coconut Water Kefir 18, 32, 34, 36, 38
Basic Ginger Bug 108, 122, 124, 126, 128
Basic Jun 82–100
Basic Kombucha 58–78
Basic Pineapple Tepache 106, 116, 118, 120
Basic Milk Kefir 20, 40, 44, 46, 142
Basic Water Kefir 16, 22, 24, 26, 28, 30
Basic Whey 142–156
basil 76
bee pollen 52
beetroot 104, 110, 112, 114
berries, mixed 86
black pepper 118
blackberries 96, 116
Blackberry Vanilla Tepache 116
blueberries 52, 92, 100
Blueberry Coconut Milk Kefir Shake 52
Breakfast Milk Kefir Shake 40
'burping' 11

C
cacao nibs 42, 44, 46
Café au Lait Milk Kefir Smoothie 44
carbonation 11

cardamom 98
Carrot Orange Kombucha 72
carrots 72
Cashew Cinnamon Coconut Milk Kefir 54
cashews 54
champagne vinegar 136
cinnamon 40, 54, 64, 74, 110, 134
cloves 64, 74
coconut 52
coconut milk 9, 20
coconut water 18
coconut yoghurt 50, 54
coffee 44
Concord Grape Shrub 132
cucumber 26, 36
Cucumber Mint Coconut Water Kefir 36

D
dates 16, 40, 42, 46

E
Earl Grey & Lavender Kombucha 68
elderflowers 94, 146
equipment 8

G
ginger 12, 24, 38, 76, 106, 108, 110, 112, 122, 124, 126, 128
Ginger Beer 122
Ginger Lime Mint Soda 124
Ginger Turmeric Coconut Water Kefir 38
grapes 132, 154
grapefruit 30, 66, 98

H
Herbal Kombucha 76
hibiscus 60, 76

Hibiscus & Lime Kombucha 60
honey 48, 50, 52, 54, 82, 146, 148, 150, 152, 154, 156

I
ice 42, 46
ingredients 8

J
jalapeño 70
jun starter liquid 82
Jun with Blackberry & Lemon Balm 96
Jun with Blueberry & Vanilla 100
Jun with Grapefruit & Cardamom 98
Jun with Jasmine Tea & Peach 84
Jun with Lavender & Blueberries 92
Jun with Mixed Berries 86
Jun with Pink Lady Apples 88
Jun with Strawberries & Thyme 90

K
kombucha starter liquid 58

L
Lacto-fermented Grape & Mint Soda 154
Lacto-fermented Orange Vanilla Soda 148
Lacto-fermented Plum & Thyme Soda 156
Lacto-fermented Rosewater 150
Lacto-fermented Summer Melon Soda 152
lavender 68, 92, 120
lemon 66, 120, 122, 146

Lemon & Lavender Pineapple
    Tepache 120
lemon balm 96
lemon verbena 62
Lemon Verbena Oolong
    Kombucha 62
lemongrass 32
lime 32, 60, 124
Lime Lemongrass Coconut
    Water Kefir 32

M
maca powder 40
mango 50
Mango Lassi Coconut Milk
    Kefir Shake 50
melon 28, 70, 152
milk 8, 20
milk kefir grains 8, 12, 20
mint 28, 36, 42, 76, 124,
    154
Mint Cacao Milk Kefir Shake
    42
molasses 144
muslin 9

O
Orange Ginger Beetroot Kvass
    112
orange 22, 66, 72, 112, 148
Orange-sicle Water Kefir 22

P
passion fruit 34
Passion Fruit Water Kefir 34
peach 84 136
Peach Champagne Shrub 136
Pear Cinnamon Shrub 134
Pear & Spice Kombucha 74
pear 74, 134
peanut butter 46
Peanut Butter & Cacao Milk
    Kefir 46
persimmon 128
Persimmon & Vanilla Ginger
    Soda 128
Piña Colada Coconut Milk
    Kefir Shake 48
pineapple 26, 48, 106, 116,
    118, 120, 126

Pineapple Black Pepper
    Tepache 118
Pineapple Cucumber Water
    Kefir 26
Pineapple Ginger Soda 126
Pink Grapefruit Water Kefir 30
plum 78, 156
Plum Darjeeling Kombucha 78
pomegranate 114
Pomegranate Beetroot Kvass
    114

R
red wine vinegar 130
resting 12
rice vinegar 138
rose petals 150
rules of fermentation 10

S
sarsaparilla bark 144
sassafras bark 144
scoby 8, 10, 11, 12, 13,
    58, 82
Spiced Beetroot Kvass 110
spirulina 42
star anise 64, 74, 110, 144
strawberries 24, 90, 130
Strawberry Ginger Water
    Kefir 24
Strawberry Shrub 130
sugar 8, 12, 16, 58, 106, 108,
    122, 124, 126, 128, 130,
    132, 134, 136, 138, 144

T
Tangerine Rice Vinegar Shrub
    138
tangerines 138
tea 9, 58, 60, 62, 64, 66, 68,
    70, 74, 76, 78, 82, 84, 86,
    90, 92, 94, 96, 98, 100
thyme 90, 156
troubleshooting 13
turmeric 38, 50

V
vanilla 22, 42, 54, 100, 116,
    128, 144, 148

W
water kefir grains 8, 12,
    16, 18
watermelon 28, 70
Watermelon Jalapeño
    Kombucha 70
Watermelon Mint Water Kefir
    28
Whey Lacto-fermented Lemon
    Soda 146
Whey Lacto-fermented Root
    Beer 144
white wine vinegar 132, 134
White Jun with Elderflowers
    94
Winter Citrus Kombucha 66

Y
yoghurt 46, 50, 54, 142

Acknowledgements

Thanks to Catie Ziller for giving me the opportunity to work on this book. Thanks to Michelle for the beautiful design. Abi Waters always makes sense of my mumbo jumbo of a manuscript and I will always be thankful. Thanks to Julia Stotz for the gorgeous photography and entertaining shoot days. I couldn't have done this without the help of Kaitlyn Kissel.

Author Biography

Caroline K. Hwang is a commercial and editorial food stylist and recipe developer. She left a career in the arts to pursue her love and passion for food and cooking by way of restaurant kitchens. She currently lives in Los Angeles, CA.

First published in 2017 by © Hachette Livre (Marabout)

The English language edition published in 2018 by Hardie Grant Books, an imprint of Hardie Grant Publishing

Hardie Grant Books (London)
5th & 6th Floors
52–54 Southwark Street
London SE1 1UN

Hardie Grant Books (Melbourne)
Building 1, 658 Church Street
Richmond, Victoria 3121

hardiegrantbooks.com

Text © Caroline Hwang
Photography © Julia Stotz

British Library Cataloguing-in-Publication Data.
A catalogue record for this book is available from the British Library.

Probiotics by Caroline Hwang

ISBN 978-1-78488-199-3

Publisher: Catie Ziller
Photography: Julia Stotz
Designer: Michelle Tilly
Editor: Abi Waters

For the English hardback edition:

Publishing Director: Kate Pollard
Junior Editor: Eila Purvis
Editor: Amy Christian
Colour Reproduction by p2d

Printed and bound in China by Leo Paper Group